51 Problems in Calculating Limits Using L'Hôpital's Rule with Solutions

by

Richard Shedenhelm

INTRODUCTION

Problems 1-23 involve the indeterminate forms $\frac{\infty}{\infty}$ and $\frac{0}{0}$. Problems 24-31 involve indeterminate products, $0 \cdot \infty$. Problems 32-40 involve indeterminate differences, $\infty - \infty$. Problems 41-51 involve indeterminate powers, 0^0, ∞^0, and 1^∞.

In the solutions, the application of L'Hôpital's Rule is denoted by "$\overset{LH}{=}$."

The appendix includes the problems stated in a random order, to help the student test his skill and detect which kinds of problems need special attention. The answers immediately following the problems in random order include the number of the original problem. Furthermore, the appendix includes a table of mathematical facts useful in calculating limits using L'Hôpital's Rule. Lastly, two applications of the rule are presented.

Athens, Georgia, April 5, 2015.

Richard Shedenhelm

PROBLEMS

Find the following limits.

1. $\lim\limits_{x\to\infty} \dfrac{e^{3x}}{x^2}$

2. $\lim\limits_{x\to 0} \dfrac{\tan(2x)}{\ln(1+x)}$

3. $\lim\limits_{x\to 0} \dfrac{\sin(x)-x}{x^3}$

4. $\lim\limits_{x\to 0} \dfrac{e^x-1}{\sin(2x)}$

5. $\lim\limits_{x\to 1} \dfrac{1-x+\ln(x)}{1+\cos(\pi x)}$

6. $\lim\limits_{\theta\to 0} \dfrac{\sin(\theta)}{\theta}$

7. $\lim\limits_{\theta\to 0} \dfrac{1-\cos(\theta)}{\theta}$

8. $\lim\limits_{t\to 0} \dfrac{1-\cos(t)}{t^2}$

9. $\lim\limits_{t\to 0^+} \dfrac{\ln(t)}{\csc(t)}$

10. $\lim\limits_{x\to 0} \dfrac{\cos(x)+2x-1}{3x}$

11. $\lim\limits_{x\to 0} \dfrac{e^x+e^{-x}-2}{1-\cos(2x)}$

12. $\lim\limits_{x\to \frac{\pi}{2}^-} \dfrac{4\tan(x)}{1+\sec(x)}$

13. $\lim\limits_{x\to\infty} \dfrac{\ln(x)}{\sqrt{x}}$

14. $\lim\limits_{x\to\infty} \dfrac{\ln(x)}{\ln(x+1)}$

15. $\lim\limits_{x\to 0} \dfrac{1-\cos(x)}{\sin(x)-x+x^2}$

16. $\lim\limits_{x\to 0} \dfrac{\sin(x)-2x}{x}$

17. $\lim\limits_{x\to\infty} \dfrac{\ln(x)}{x^{10}}$

18. $\lim\limits_{x\to 0} \dfrac{10^x-2^x}{6x}$

19. $\lim\limits_{x\to 0} \dfrac{x^2}{\ln(\cos(x))}$

20. $\lim\limits_{x\to 0} \dfrac{x\cdot 2^x}{2^x-1}$

21. $\lim\limits_{x\to 0^+} \dfrac{\ln(x^2+2x)}{\ln(x)}$

22. $\lim\limits_{x\to\infty} \dfrac{e^x+x^2}{e^x-x}$

23. $\lim\limits_{x\to 0} \dfrac{2\cos(\theta)-2}{e^\theta-\theta-1}$

24. $\lim\limits_{\theta\to\infty} x^2 e^{-x}$

25. $\lim\limits_{x\to 0^+} x\ln(x)$

26. $\lim\limits_{x\to 0^+} x^2\ln(x)$

27. $\lim\limits_{x\to \frac{\pi}{2}} (2x-\pi)\sec(x)$

28. $\lim\limits_{x\to \frac{\pi}{2}} \tan(x)\ln(\sin(x))$

29. $\lim\limits_{x\to \frac{\pi}{2}} \left(x-\dfrac{\pi}{2}\right)\sec(x)$

30. $\lim\limits_{x\to 0^+} \sin(x)\ln(\sin(x))$

31. $\lim\limits_{x\to 0} x\cot(x)$

32. $\lim\limits_{x\to\infty} [\ln(2x)-\ln(x+1)]$

33. $\lim\limits_{x\to \frac{\pi}{2}} (\sec(x)-\tan(x))$

34. $\lim\limits_{x\to 0} \left(\dfrac{1}{e^x-1}-\dfrac{1}{x}\right)$

35. $\lim\limits_{x\to 1} \left(\dfrac{x}{x-1}-\dfrac{1}{\ln(x)}\right)$

36. $\lim\limits_{x\to 0} \left(\dfrac{1}{x}-\dfrac{1}{\sin(x)}\right)$

37. $\lim\limits_{x\to\infty} \left(\sqrt{x^2+3x}-x\right)$

38. $\lim\limits_{x\to 1^+} \left(\dfrac{1}{x-1}-\dfrac{1}{\ln(x)}\right)$

39. $\lim\limits_{x\to\infty} \left(\dfrac{x^2}{x-1}-\dfrac{x^2}{x+1}\right)$

40. $\lim\limits_{x\to\infty} \left(\sqrt{x^2+x}-\sqrt{x^2-x}\right)$

41. $\lim\limits_{x\to 0^+} (1+\sin(4x))^{\cot(x)}$

42. $\lim\limits_{x\to 0^+} x^x$

43. $\lim\limits_{x\to 0} (1+x)^{\frac{1}{x}}$

44. $\lim\limits_{x\to 1^+} x^{\frac{1}{1-x}}$

45. $\lim\limits_{x\to\infty} [\ln(x)]^{\frac{1}{x}}$

46. $\lim\limits_{x\to 0^+} x^{\frac{-1}{\ln(x)}}$

47. $\lim\limits_{x\to\infty} (1+2x)^{\frac{1}{2\ln(x)}}$

48. $\lim\limits_{x\to 0} (e^x+x)^{\frac{2}{x}}$

49. $\lim\limits_{h\to 0} (1+hx)^{\frac{1}{h}}$

50. $\lim\limits_{n\to\infty} \left(1+\dfrac{x}{n}\right)^n$

51. $\lim\limits_{n\to\infty} \left(1+\dfrac{r}{n}\right)^{nt}$

5

ANSWERS

1. ∞

2. 2

3. $\frac{-1}{6}$

4. $\frac{1}{2}$

5. $\frac{-1}{\pi^2}$

6. 1

7. 0

8. $\frac{1}{2}$

9. 0

10. $\frac{2}{3}$

11. $\frac{1}{2}$

12. 4

13. 0

14. 1

15. $\frac{1}{2}$

16. -1

17. 0

18. $\frac{\ln(5)}{6}$

19. -2

20. $\frac{1}{\ln(2)}$

21. 1

22. 1

23. -2

24. 0

25. 0

26. 0

27. -2

28. 0

29. 1

30. 0

31. 1

32. $\ln(2)$

33. 0

34. $\frac{-1}{2}$

35. $\frac{1}{2}$

36. 0

37. $\frac{3}{2}$

38. $\frac{-1}{2}$

39. 2

40. 1

41. e^4

42. 1

43. e

44. $\frac{1}{e}$

45. 1

46. $\frac{1}{e}$

47. $e^{\frac{1}{2}}$

48. e^4

49. e^x

50. e^x

51. e^{rt}

SOLUTIONS

1. $\displaystyle\lim_{x\to\infty}\frac{e^{3x}}{x^2}\overset{LH}{=}\lim_{x\to\infty}\frac{3e^{3x}}{2x}\overset{LH}{=}\lim_{x\to\infty}\frac{9e^{3x}}{2}=\infty.$

2. $\displaystyle\lim_{x\to0}\frac{\tan(2x)}{\ln(1+x)}\overset{LH}{=}\lim_{x\to0}\frac{2\sec^2(2x)}{\dfrac{1}{1+x}}=\lim_{x\to0}\frac{2\sec^2(2x)}{1}\cdot\frac{1+x}{1}=\frac{2\sec^2(0)}{1}\cdot\frac{1+0}{1}=2\cdot1=2.$

3. $\displaystyle\lim_{x\to0}\frac{\sin(x)-x}{x^3}\overset{LH}{=}\lim_{x\to0}\frac{\cos(x)-1}{3x^2}\overset{LH}{=}\lim_{x\to0}\frac{-\sin(x)}{6x}\overset{LH}{=}\lim_{x\to0}\frac{-\cos(x)}{6}=\frac{-\cos(0)}{6}=-\frac{1}{6}.$

4. $\displaystyle\lim_{x\to0}\frac{e^x-1}{\sin(2x)}\overset{LH}{=}\lim_{x\to0}\frac{e^x}{2\cos(2x)}=\frac{e^0}{2\cos(0)}=\frac{1}{2(1)}=\frac{1}{2}.$

5. $\displaystyle\lim_{x\to1}\frac{1-x+\ln(x)}{1+\cos(\pi x)}\overset{LH}{=}\lim_{x\to1}\frac{-1+\dfrac{1}{x}}{-\pi\sin(\pi x)}=\lim_{x\to1}\frac{x-1}{\pi x\sin(\pi x)}\overset{LH}{=}\lim_{x\to1}\frac{1}{\pi\sin(\pi x)+\pi^2 x\cos(\pi x)}=$

   $=\dfrac{1}{\pi\sin(\pi)+\pi^2\cos(\pi)}=\dfrac{1}{\pi(0)+\pi^2(-1)}=\dfrac{-1}{\pi^2}.$

6. $\displaystyle\lim_{\theta\to0}\frac{\sin(\theta)}{\theta}\overset{LH}{=}\lim_{\theta\to0}\frac{\cos(\theta)}{1}=\frac{\cos(0)}{1}=\frac{1}{1}=1.$

7. $\displaystyle\lim_{\theta\to0}\frac{1-\cos(\theta)}{\theta}\overset{LH}{=}\lim_{\theta\to0}\frac{\sin(\theta)}{1}=\frac{\sin(0)}{1}=\frac{0}{1}=0.$

8. $\displaystyle\lim_{t\to0}\frac{1-\cos(t)}{t^2}\overset{LH}{=}\lim_{t\to0}\frac{\sin(t)}{2t}\overset{LH}{=}\lim_{t\to0}\frac{\cos(t)}{2}=\frac{\cos(0)}{2}=\frac{1}{2}.$

9. $$\lim_{t\to 0^+} \frac{\ln(t)}{\csc(t)} \overset{LH}{=} \lim_{t\to 0^+} \frac{\frac{1}{t}}{-\csc(t)\cot(t)} = \lim_{t\to 0^+} -\frac{1}{t} \cdot \frac{1}{\csc(t)\cot(t)} = \lim_{t\to 0^+} -\frac{1}{t} \cdot \frac{1}{\frac{1}{\sin(t)} \cdot \frac{\cos(t)}{\sin(t)}} =$$

$$= \lim_{t\to 0^+} -\frac{1}{t} \cdot \frac{\sin^2(t)}{\cos(t)} = \lim_{t\to 0^+} -\frac{\sin(t)\tan(t)}{t} \overset{LH}{=} \lim_{t\to 0^+} -\frac{\cos(t)\tan(t) + \sin(t)\sec^2(t)}{1} =$$

$$= -\frac{\cos(0)\tan(0) + \sin(0)\sec^2(0)}{1} = -\frac{(1)(0) + (0)(1)}{1} = -\frac{0+0}{1} = 0.$$

10. $$\lim_{x\to 0} \frac{\cos(x) + 2x - 1}{3x} \overset{LH}{=} \lim_{x\to 0} \frac{-\sin(x) + 2}{3} = \frac{-\sin(0) + 2}{3} = \frac{-0+2}{3} = \frac{2}{3}.$$

11. $$\lim_{x\to 0} \frac{e^x + e^{-x} - 2}{1 - \cos(2x)} \overset{LH}{=} \lim_{x\to 0} \frac{e^x - e^{-x}}{2\sin(2x)} \overset{LH}{=} \lim_{x\to 0} \frac{e^x + e^{-x}}{4\cos(2x)} = \frac{e^0 + e^{-0}}{4\cos(0)} = \frac{1+1}{4(1)} = \frac{2}{4} = \frac{1}{2}.$$

12. $$\lim_{x\to \frac{\pi}{2}^-} \frac{4\tan(x)}{1 + \sec(x)} \overset{LH}{=} \lim_{x\to \frac{\pi}{2}^-} \frac{4\sec^2(x)}{\sec(x)\tan(x)} = \lim_{x\to \frac{\pi}{2}^-} \frac{4\sec(x)}{\tan(x)} = \lim_{x\to \frac{\pi}{2}^-} \frac{\left(\frac{4}{\cos(x)}\right)}{\left(\frac{\sin(x)}{\cos(x)}\right)} =$$

$$= \lim_{x\to \frac{\pi}{2}^-} \frac{4}{\sin(x)} = \frac{4}{\sin\left(\frac{\pi}{2}\right)} = \frac{4}{1} = 4.$$

13. $$\lim_{x\to \infty} \frac{\ln(x)}{\sqrt{x}} = \lim_{x\to \infty} \frac{\ln(x)}{x^{\frac{1}{2}}} \overset{LH}{=} \lim_{x\to \infty} \frac{\left(\frac{1}{x}\right)}{\frac{1}{2}x^{-\frac{1}{2}}} = \lim_{x\to \infty} \frac{\left(\frac{1}{x}\right)}{\left(\frac{1}{2x^{\frac{1}{2}}}\right)} = \lim_{x\to \infty} \frac{\left(\frac{1}{x}\right)}{\left(\frac{1}{2\sqrt{x}}\right)} = \lim_{x\to \infty} \frac{1}{x} \cdot \frac{2\sqrt{x}}{1} =$$

$$= \lim_{x\to \infty} \frac{2}{\sqrt{x}} = 0.$$

14. $$\lim_{x\to \infty} \frac{\ln(x)}{\ln(x+1)} \overset{LH}{=} \lim_{x\to \infty} \frac{\left(\frac{1}{x}\right)}{\left(\frac{1}{x+1}\right)} = \lim_{x\to \infty} \frac{x+1}{x} \overset{LH}{=} \lim_{x\to \infty} \frac{1}{1} = 1.$$

15. $$\lim_{x\to 0} \frac{1 - \cos(x)}{\sin(x) - x + x^2} \overset{LH}{=} \lim_{x\to 0} \frac{\sin(x)}{\cos(x) - 1 + 2x} \overset{LH}{=} \lim_{x\to 0} \frac{\cos(x)}{-\sin(x) + 2} = \frac{\cos(0)}{-\sin(0) + 2} = \frac{1}{-0 + 2} =$$

$$= \frac{1}{2}.$$

16. $$\lim_{x\to 0}\frac{\sin(x)-2x}{x}\overset{LH}{=}\lim_{x\to 0}\frac{\cos(x)-2}{1}=\frac{\cos(0)-2}{1}=\frac{1-2}{1}=-1.$$

17. $$\lim_{x\to\infty}\frac{\ln(x)}{x^{10}}\overset{LH}{=}\lim_{x\to\infty}\frac{\left(\frac{1}{x}\right)}{10x^9}=\lim_{x\to\infty}\frac{1}{10x^{10}}=0.$$

18. $$\lim_{x\to 0}\frac{10^x-2^x}{6x}\overset{LH}{=}\lim_{x\to 0}\frac{\ln(10)\,10^x-\ln(2)\,2^x}{6}=\frac{\ln(10)\,10^0-\ln(2)\,2^0}{6}=\frac{\ln(10)-\ln(2)}{6}=$$
$$=\frac{\ln\left(\frac{10}{2}\right)}{6}=\frac{\ln(5)}{6}.$$

19. $$\lim_{x\to 0}\frac{x^2}{\ln(\cos(x))}\overset{LH}{=}\lim_{x\to 0}\frac{2x}{\left(\frac{-\sin(x)}{\cos(x)}\right)}=\lim_{x\to 0}\frac{-2x\cos(x)}{\sin(x)}\overset{LH}{=}\lim_{x\to 0}\frac{-2\cos(x)+2x\sin(x)}{\cos(x)}=$$
$$=\frac{-2\cos(0)+2(0)\sin(0)}{\cos(0)}=\frac{-2(1)+2(0)(0)}{1}=\frac{-2}{1}=-2.$$

20. $$\lim_{x\to 0}\frac{x\cdot 2^x}{2^x-1}\overset{LH}{=}\lim_{x\to 0}\frac{2^x+x\ln(2)2^x}{\ln(2)\,2^x}=\lim_{x\to 0}\frac{2^x(1+x\ln(2))}{\ln(2)\,2^x}=\lim_{x\to 0}\frac{1+x\ln(2)}{\ln(2)}=\frac{1+(0)\ln(2)}{\ln(2)}=$$
$$=\frac{1}{\ln(2)}.$$

21. $$\lim_{x\to 0^+}\frac{\ln(x^2+2x)}{\ln(x)}\overset{LH}{=}\lim_{x\to 0^+}\frac{\left(\frac{2x+2}{x^2+2x}\right)}{\left(\frac{1}{x}\right)}=\lim_{x\to 0^+}\frac{2x+2}{x^2+2x}\cdot\frac{x}{1}=\lim_{x\to 0^+}\frac{2(x+1)x}{(x+2)x}=$$
$$=\lim_{x\to 0^+}\frac{2(x+1)}{(x+2)}=\frac{2(0+1)}{0+2}=\frac{2}{2}=1.$$

22. $$\lim_{x\to\infty}\frac{e^x+x^2}{e^x-x}\overset{LH}{=}\lim_{x\to\infty}\frac{e^x+2x}{e^x-1}\overset{LH}{=}\lim_{x\to\infty}\frac{e^x+2}{e^x}\overset{LH}{=}\lim_{x\to\infty}\frac{e^x}{e^x}=\lim_{x\to\infty}1=1.$$

23. $$\lim_{\theta\to 0}\frac{2\cos(\theta)-2}{e^\theta-\theta-1}\overset{LH}{=}\lim_{\theta\to 0}\frac{-2\sin(\theta)}{e^\theta-1}\overset{LH}{=}\lim_{\theta\to 0}\frac{-2\cos(\theta)}{e^\theta}=\frac{-2\cos(0)}{e^0}=\frac{-2(1)}{1}=\frac{-2}{1}=-2.$$

24. $$\lim_{x\to\infty} x^2 e^{-x} = \lim_{x\to\infty} \frac{x^2}{e^x} \stackrel{LH}{=} \lim_{x\to\infty} \frac{2x}{e^x} \stackrel{LH}{=} \lim_{x\to\infty} \frac{2}{e^x} = 0.$$

25. $$\lim_{x\to 0^+} x\ln(x) = \lim_{x\to 0^+} \frac{\ln(x)}{\left(\frac{1}{x}\right)} = \lim_{x\to 0^+} \frac{\ln(x)}{x^{-1}} \stackrel{LH}{=} \lim_{x\to 0^+} \frac{\frac{1}{x}}{-x^{-2}} = \lim_{x\to 0^+} \frac{\left(\frac{1}{x}\right)}{\left(\frac{-1}{x^2}\right)} = \lim_{x\to 0^+} \frac{1}{x} \cdot \frac{x^2}{-1} =$$
$$= -\lim_{x\to 0^+} x = -0 = 0.$$

26. $$\lim_{x\to 0^+} x^2 \ln(x) = \lim_{x\to 0^+} \frac{\ln(x)}{\left(\frac{1}{x^2}\right)} = \lim_{x\to 0^+} \frac{\ln(x)}{x^{-2}} \stackrel{LH}{=} \lim_{x\to 0^+} \frac{\frac{1}{x}}{-2x^{-3}} = \lim_{x\to 0^+} \frac{\left(\frac{1}{x}\right)}{\left(\frac{-2}{x^3}\right)} = \lim_{x\to 0^+} \frac{1}{x} \cdot \frac{x^3}{-2} =$$
$$= \frac{-1}{2} \lim_{x\to 0^+} x^2 = \frac{-1}{2}(0)^2 = 0.$$

27. $$\lim_{x\to\frac{\pi}{2}} (2x - \pi)\sec(x) = \lim_{x\to\frac{\pi}{2}} \frac{2x - \pi}{\cos(x)} \stackrel{LH}{=} \lim_{x\to\frac{\pi}{2}} \frac{2}{-\sin(x)} = \frac{2}{-\sin\left(\frac{\pi}{2}\right)} = \frac{2}{-1} = -2.$$

28. $$\lim_{x\to\frac{\pi}{2}} \tan(x)\ln(\sin(x)) = \lim_{x\to\frac{\pi}{2}} \frac{\ln(\sin(x))}{\cot(x)} \stackrel{LH}{=} \lim_{x\to\frac{\pi}{2}} \frac{\left(\frac{\cos(x)}{\sin(x)}\right)}{-\csc^2(x)} = \lim_{x\to\frac{\pi}{2}} \frac{\cos(x)}{\sin(x)} \cdot \frac{-1}{\csc^2(x)} =$$
$$= \lim_{x\to\frac{\pi}{2}} \frac{\cos(x)}{\sin(x)} \cdot \frac{-\sin^2(x)}{1} = -\lim_{x\to\frac{\pi}{2}} \cos(x)\sin(x) = -\cos\left(\frac{\pi}{2}\right)\sin\left(\frac{\pi}{2}\right) = -(0)(1) = 0.$$

29. $$\lim_{x\to\frac{\pi}{2}} \left(x - \frac{\pi}{2}\right)\sec(x) = \lim_{x\to\frac{\pi}{2}} \frac{x - \frac{\pi}{2}}{\cos(x)} \stackrel{LH}{=} \lim_{x\to\frac{\pi}{2}} \frac{1}{-\sin(x)} = \frac{1}{-\sin\left(\frac{\pi}{2}\right)} = \frac{1}{-1} = 1.$$

30. $$\lim_{x\to 0^+} \sin(x)\ln(\sin(x)) = \lim_{x\to 0^+} \frac{\ln(\sin(x))}{\frac{1}{\sin(x)}} \stackrel{LH}{=} \lim_{x\to 0^+} \frac{\left(\frac{\cos(x)}{\sin(x)}\right)}{\left(\frac{-\cos(x)}{\sin^2(x)}\right)} = -\lim_{x\to 0^+} \frac{\cos(x)}{\sin(x)} \cdot \frac{\sin^2(x)}{\cos(x)} =$$
$$= -\lim_{x\to 0^+} \sin(x) = -\sin(0) = 0.$$

31. $\lim\limits_{x\to 0} x\cot(x) = \lim\limits_{x\to 0}\dfrac{x}{\tan(x)} \overset{LH}{=} \lim\limits_{x\to 0}\dfrac{1}{\sec^2(x)} = \lim\limits_{x\to 0}\cos^2(x) = \cos^2(0) = 1.$

32. $\lim\limits_{x\to\infty}[\ln(2x) - \ln(x+1)] = \lim\limits_{x\to\infty}\ln\left(\dfrac{2x}{x+1}\right) = \ln\left(\lim\limits_{x\to\infty}\dfrac{2x}{x+1}\right) \overset{LH}{=} \ln\left(\lim\limits_{x\to\infty}\dfrac{2}{1}\right) = \ln\left(\lim\limits_{x\to\infty}2\right) =$

$= \ln(2).$

33. $\lim\limits_{x\to\frac{\pi}{2}^-}(\sec(x) - \tan(x)) = \lim\limits_{x\to\frac{\pi}{2}^-}\left(\dfrac{1}{\cos(x)} - \dfrac{\sin(x)}{\cos(x)}\right) = \lim\limits_{x\to\frac{\pi}{2}^-}\dfrac{1 - \sin(x)}{\cos(x)} \overset{LH}{=} \lim\limits_{x\to\frac{\pi}{2}^-}\dfrac{-\cos(x)}{-\sin(x)} =$

$= \dfrac{\cos\left(\frac{\pi}{2}\right)}{\sin\left(\frac{\pi}{2}\right)} = \dfrac{0}{1} = 0.$

34. $\lim\limits_{x\to 0}\left(\dfrac{1}{e^x - 1} - \dfrac{1}{x}\right) = \lim\limits_{x\to 0}\dfrac{x - e^x + 1}{xe^x - x} \overset{LH}{=} \lim\limits_{x\to 0}\dfrac{1 - e^x}{xe^x + e^x - 1} \overset{LH}{=} \lim\limits_{x\to 0}\dfrac{-e^x}{xe^x + 2e^x} = \dfrac{-e^0}{(0)e^0 + 2e^0} =$

$= \dfrac{-1}{0+2} = \dfrac{-1}{2}.$

35. $\lim\limits_{x\to 1}\left(\dfrac{x}{x-1} - \dfrac{1}{\ln(x)}\right) = \lim\limits_{x\to 1}\dfrac{x\ln(x) - (x+1)}{(x-1)\ln(x)} = \lim\limits_{x\to 1}\dfrac{x\ln(x) - x - 1}{x\ln(x) - \ln(x)} \overset{LH}{=} \lim\limits_{x\to 1}\dfrac{\ln(x) + 1 - 1}{\ln(x) + 1 - \frac{1}{x}} =$

$= \lim\limits_{x\to 1}\dfrac{\ln(x)}{\left(\dfrac{x\ln(x) + x - 1}{x}\right)} = \lim\limits_{x\to 1}\dfrac{x\ln(x)}{x\ln(x) + x - 1} \overset{LH}{=} \lim\limits_{x\to 1}\dfrac{\ln(x) + 1}{\ln(x) + 1 + 1} = \lim\limits_{x\to 1}\dfrac{\ln(x) + 1}{\ln(x) + 2} =$

$= \dfrac{\ln(1) + 1}{\ln(1) + 2} = \dfrac{0+1}{0+2} = \dfrac{1}{2}.$

36. $\lim\limits_{x\to 0}\left(\dfrac{1}{x} - \dfrac{1}{\sin(x)}\right) = \lim\limits_{x\to 0}\dfrac{\sin(x) - x}{x\sin(x)} \overset{LH}{=} \lim\limits_{x\to 0}\dfrac{\cos(x) - 1}{\sin(x) + x\cos(x)} \overset{LH}{=}$

$\overset{LH}{=} \lim\limits_{x\to 0}\dfrac{-\sin(x)}{\cos(x) + \cos(x) - x\sin(x)} = \lim\limits_{x\to 0}\dfrac{-\sin(x)}{2\cos(x) - x\sin(x)} = \dfrac{-\sin(0)}{2\cos(0) - (0)\sin(0)} =$

$= \dfrac{0}{2(1) - 0} = \dfrac{0}{2} = 0.$

37. $\displaystyle\lim_{x\to\infty}\left(\sqrt{x^2+3x}-x\right)=\lim_{x\to\infty}\left(\sqrt{x^2\left(1+\tfrac{3}{x}\right)}-x\right)=\lim_{x\to\infty}\left(x\sqrt{1+\tfrac{3}{x}}-x\right)=$

$\displaystyle=\lim_{x\to\infty}x\left(\sqrt{1+\tfrac{3}{x}}-1\right)=\lim_{x\to\infty}\frac{\sqrt{1+\tfrac{3}{x}}-1}{\left(\tfrac{1}{x}\right)}=\lim_{x\to\infty}\frac{\left(1+\tfrac{3}{x}\right)^{\frac{1}{2}}-1}{\left(\tfrac{1}{x}\right)}=\lim_{x\to\infty}\frac{(1+3x^{-1})^{\frac{1}{2}}-1}{x^{-1}}\overset{LH}{\triangleq}$

$\displaystyle\overset{LH}{\triangleq}\lim_{x\to\infty}\frac{\tfrac{1}{2}(1+3x^{-1})^{\frac{-1}{2}}(-3x^{-2})}{-x^{-2}}=\lim_{x\to\infty}\frac{\left(\tfrac{1}{2(1+\tfrac{3}{x})^{\frac{1}{2}}}\right)\cdot\left(\tfrac{-3}{x^2}\right)}{\left(\tfrac{-1}{x^2}\right)}=\lim_{x\to\infty}\frac{\left(\tfrac{-3}{2x^2\sqrt{1+\tfrac{3}{x}}}\right)}{\left(\tfrac{-1}{x^2}\right)}=$

$\displaystyle=\lim_{x\to\infty}\left(\frac{-3}{2x^2\sqrt{1+\tfrac{3}{x}}}\right)\cdot\left(\frac{x^2}{-1}\right)=\lim_{x\to\infty}\frac{3}{2\sqrt{1+\tfrac{3}{x}}}=\frac{3}{2\sqrt{1+0}}=\frac{3}{2}.$

38. $\displaystyle\lim_{x\to1^+}\left(\frac{1}{x-1}-\frac{1}{\ln(x)}\right)=\lim_{x\to1^+}\frac{\ln(x)-(x-1)}{\ln(x)(x-1)}=\lim_{x\to1^+}\frac{\ln(x)-x+1}{\ln(x)(x-1)}\overset{LH}{\triangleq}$

$\displaystyle\overset{LH}{\triangleq}\lim_{x\to1^+}\frac{\tfrac{1}{x}-1}{\tfrac{1}{x}(x-1)+\ln(x)}=\lim_{x\to1^+}\frac{\left(\tfrac{1-x}{x}\right)}{\left(\tfrac{x-1+x\ln(x)}{x}\right)}=\lim_{x\to1^+}\frac{1-x}{x}\cdot\frac{x}{x-1+x\ln(x)}=$

$\displaystyle=\lim_{x\to1^+}\frac{1-x}{x-1+x\ln(x)}\overset{LH}{\triangleq}\lim_{x\to1^+}\frac{-1}{1+\ln(x)+1}=\frac{-1}{2+\ln(1)}=\frac{-1}{2+0}=\frac{-1}{2}.$

39. $\displaystyle\lim_{x\to\infty}\left(\frac{x^2}{x-1}-\frac{x^2}{x+1}\right)=\lim_{x\to\infty}\frac{x^2(x+1)-x^2(x-1)}{(x-1)(x+1)}=\lim_{x\to\infty}\frac{x^3+x^2-x^3+x^2}{x^2-1}=$

$\displaystyle=\lim_{x\to\infty}\frac{2x^2}{x^2-1}\overset{LH}{\triangleq}\lim_{x\to\infty}\frac{4x}{2x}=\lim_{x\to\infty}2=2.$

12

40. $\displaystyle\lim_{x\to\infty}\left(\sqrt{x^2+x}-\sqrt{x^2-x}\right)=\lim_{x\to\infty}\frac{\sqrt{x^2+x}-\sqrt{x^2-x}}{1}\cdot\frac{\sqrt{x^2+x}+\sqrt{x^2-x}}{\sqrt{x^2+x}+\sqrt{x^2-x}}=$

$\displaystyle=\lim_{x\to\infty}\frac{(x^2+x)-(x^2-x)}{\sqrt{x^2+x}+\sqrt{x^2-x}}=\lim_{x\to\infty}\frac{x^2+x-x^2+x}{\sqrt{x^2+x}+\sqrt{x^2-x}}=\lim_{x\to\infty}\frac{2x}{(x^2+x)^{\frac12}+(x^2-x)^{\frac12}}\overset{LH}{=\!=}$

$\displaystyle\overset{LH}{=\!=}\lim_{x\to\infty}\frac{2}{\frac12(x^2+x)^{\frac{-1}{2}}(2x+1)+\frac12(x^2-x)^{\frac{-1}{2}}(2x-1)}=\lim_{x\to\infty}\frac{2}{\dfrac{2x+1}{2\sqrt{x^2+x}}+\dfrac{2x-1}{2\sqrt{x^2-x}}}=$

$\displaystyle=\lim_{x\to\infty}\frac{2}{\dfrac{x\left(2+\frac1x\right)}{2\sqrt{x^2\left(1+\frac1x\right)}}+\dfrac{x\left(2-\frac1x\right)}{2\sqrt{x^2\left(1-\frac1x\right)}}}=\lim_{x\to\infty}\frac{2}{\dfrac{x\left(2+\frac1x\right)}{2x\sqrt{1+\frac1x}}+\dfrac{x\left(2-\frac1x\right)}{2x\sqrt{1-\frac1x}}}=$

$\displaystyle=\lim_{x\to\infty}\frac{2}{\dfrac{2+\frac1x}{2\sqrt{1+\frac1x}}+\dfrac{2-\frac1x}{2\sqrt{1-\frac1x}}}=\frac{2}{\dfrac{2+0}{2\sqrt{1+0}}+\dfrac{2-0}{2\sqrt{1-0}}}=\frac{2}{\dfrac{2}{2\sqrt1}+\dfrac{2}{2\sqrt1}}=\frac{2}{\frac22+\frac22}=\frac{2}{1+1}=\frac22=1.$

41. Let $y=(1+\sin(4x))^{\cot(x)}$.

Hence, $\ln(y)=\ln\left[(1+\sin(4x))^{\cot(x)}\right]=\cot(x)\ln(1+\sin(4x))$.

So, $\displaystyle\lim_{x\to0^+}\ln(y)=\lim_{x\to0^+}\cot(x)\ln(1+\sin(4x))=\lim_{x\to0^+}\frac{\ln(1+\sin(4x))}{\tan(x)}\overset{LH}{=\!=}\lim_{x\to0^+}\frac{\left(\dfrac{4\cos(x)}{1+\sin(4x)}\right)}{\sec^2(x)}=$

$\displaystyle=\frac{\left(\dfrac{4\cos(0)}{1+\sin(0)}\right)}{\sec^2(0)}=\frac{\left(\dfrac{4(1)}{1+0}\right)}{1}=\frac{\left(\frac41\right)}{1}=4.$ That is, $\displaystyle\lim_{x\to0^+}\ln(y)=4.$

Now $y=y\Leftrightarrow\ln(y)=\ln(y)\Leftrightarrow e^{\ln(y)}=y$. Therefore, $\displaystyle\lim_{x\to0^+}(1+\sin(4x))^{\cot(x)}=\lim_{x\to0^+}y=$

$\displaystyle=\lim_{x\to0^+}e^{\ln(y)}=e^{\lim_{x\to0^+}\ln(y)}=e^4.$

42. Let $y = x^x$.

Hence, $\ln(y) = \ln(x^x) = x \ln(x)$.

So, $\lim\limits_{x \to 0^+} \ln(y) = \lim\limits_{x \to 0^+} x \ln(x) = \lim\limits_{x \to 0^+} \dfrac{\ln(x)}{\left(\frac{1}{x}\right)} = \lim\limits_{x \to 0^+} \dfrac{\ln(x)}{x^{-1}} \overset{LH}{=} \lim\limits_{x \to 0^+} \dfrac{\left(\frac{1}{x}\right)}{-x^{-2}} = \lim\limits_{x \to 0^+} \dfrac{\left(\frac{1}{x}\right)}{\left(\frac{-1}{x^2}\right)} =$

$= \lim\limits_{x \to 0^+} \dfrac{1}{x} \cdot \dfrac{x^2}{-1} = \lim\limits_{x \to 0^+} -x = 0$. That is, $\lim\limits_{x \to 0^+} \ln(y) = 0$.

Now $y = y \iff \ln(y) = \ln(y) \iff e^{\ln(y)} = y$. Therefore, $\lim\limits_{x \to 0^+} x^x = \lim\limits_{x \to 0^+} y =$

$= \lim\limits_{x \to 0^+} e^{\ln(y)} = e^{\lim\limits_{x \to 0^+} \ln(y)} = e^0 = 1$.

43. Let $y = (1 + x)^{\frac{1}{x}}$.

Hence, $\ln(y) = \ln\left[(1 + x)^{\frac{1}{x}}\right] = \frac{1}{x}\ln(1 + x) = \dfrac{\ln(1 + x)}{x}$.

So, $\lim\limits_{x \to 0} \ln(y) = \lim\limits_{x \to 0} \dfrac{\ln(1+x)}{x} \overset{LH}{=} \lim\limits_{x \to 0} \dfrac{1}{1+x} = \dfrac{1}{1+0} = 1$. That is, $\lim\limits_{x \to 0} \ln(y) = 1$.

Now $y = y \iff \ln(y) = \ln(y) \iff e^{\ln(y)} = y$. Therefore, $\lim\limits_{x \to 0}(1 + x)^{\frac{1}{x}} = \lim\limits_{x \to 0} y =$

$= \lim\limits_{x \to 0} e^{\ln(y)} = e^{\lim\limits_{x \to 0} \ln(y)} = e^1 = e$.

44. Let $y = x^{\frac{1}{1-x}}$.

Hence, $\ln(y) = \ln\left(x^{\frac{1}{1-x}}\right) = \frac{1}{1-x}\ln(x) = \dfrac{\ln(x)}{1 - x}$.

So, $\lim\limits_{x \to 1^+} \ln(y) = \lim\limits_{x \to 1^+} \dfrac{\ln(x)}{1-x} \overset{LH}{=} \lim\limits_{x \to 1^+} \dfrac{\left(\frac{1}{x}\right)}{-1} = \dfrac{\left(\frac{1}{1}\right)}{-1} = -1$. That is, $\lim\limits_{x \to 1^+} \ln(y) = -1$.

Now $y = y \iff \ln(y) = \ln(y) \iff e^{\ln(y)} = y$. Therefore, $\lim\limits_{x \to 1^+} x^{\frac{1}{1-x}} = \lim\limits_{x \to 1^+} y =$

$= \lim\limits_{x \to 1^+} e^{\ln(y)} = e^{\lim\limits_{x \to 1^+} \ln(y)} = e^{-1} = \dfrac{1}{e}$.

45. Let $y = [\ln(x)]^{\frac{1}{x}}$.

Hence, $\ln(y) = \ln\left[[\ln(x)]^{\frac{1}{x}}\right] = \frac{1}{x}\ln[\ln(x)] = \frac{\ln[\ln(x)]}{x}$.

So, $\lim\limits_{x\to\infty} \ln(y) = \lim\limits_{x\to\infty} \frac{\ln[\ln(x)]}{x} \overset{LH}{=} \lim\limits_{x\to\infty} \frac{\frac{1}{\ln(x)}\frac{1}{x}}{1} = \frac{0}{1} = 0$. That is, $\lim\limits_{x\to\infty} \ln(y) = 0$.

Now $y = y \Leftrightarrow \ln(y) = \ln(y) \Leftrightarrow e^{\ln(y)} = y$. Therefore, $\lim\limits_{x\to\infty} [\ln(x)]^{\frac{1}{x}} = \lim\limits_{x\to\infty} y =$

$= \lim\limits_{x\to\infty} e^{\ln(y)} = e^{\lim\limits_{x\to\infty}\ln(y)} = e^0 = 1$.

46. Let $y = x^{\frac{-1}{\ln(x)}}$.

Hence, $\ln(y) = \ln\left(x^{\frac{-1}{\ln(x)}}\right) = \frac{-1}{\ln(x)}\ln(x) = \frac{-\ln(x)}{\ln(x)} = -1$.

So, $\lim\limits_{x\to 0^+} \ln(y) = \lim\limits_{x\to 0^+} (-1) = -1$.

Now $y = y \Leftrightarrow \ln(y) = \ln(y) \Leftrightarrow e^{\ln(y)} = y$. Therefore, $\lim\limits_{x\to 0^+} x^{\frac{-1}{\ln(x)}} = \lim\limits_{x\to 0^+} y =$

$= \lim\limits_{x\to 0^+} e^{\ln(y)} = e^{\lim\limits_{x\to 0^+}\ln(y)} = e^{-1} = \frac{1}{e}$.

47. Let $y = (1 + 2x)^{\frac{1}{2\ln(x)}}$.

Hence, $\ln(y) = \ln\left[(1 + 2x)^{\frac{1}{2\ln(x)}}\right] = \frac{1}{2\ln(x)}\ln(1 + 2x) = \frac{\ln(1 + 2x)}{2\ln(x)}$.

So, $\lim\limits_{x\to\infty} \ln(y) = \lim\limits_{x\to\infty} \frac{\ln(1+2x)}{2\ln(x)} \overset{LH}{=} \lim\limits_{x\to\infty} \frac{\left(\frac{2}{1+2x}\right)}{\left(\frac{2}{x}\right)} = \lim\limits_{x\to\infty} \frac{2}{1+2x}\cdot\frac{x}{2} = \lim\limits_{x\to\infty} \frac{x}{1+2x} \overset{LH}{=} \lim\limits_{x\to\infty} \frac{1}{2} = \frac{1}{2}$.

That is, $\lim\limits_{x\to\infty} \ln(y) = \frac{1}{2}$.

Now $y = y \Leftrightarrow \ln(y) = \ln(y) \Leftrightarrow e^{\ln(y)} = y$. Therefore, $\lim\limits_{x\to\infty} (1 + 2x)^{\frac{1}{2\ln(x)}} = \lim\limits_{x\to\infty} y =$

$= \lim\limits_{x\to\infty} e^{\ln(y)} = e^{\lim\limits_{x\to\infty}\ln(y)} = e^{\frac{1}{2}}$.

48. Let $y = (e^x + x)^{\frac{2}{x}}$.

Hence, $\ln(y) = \ln\left[(e^x + x)^{\frac{2}{x}}\right] = \frac{2}{x}\ln(e^x + x) = \frac{2\ln(e^x + x)}{x}$.

So, $\lim\limits_{x \to 0} \ln(y) = \lim\limits_{x \to 0} \frac{2\ln(e^x + x)}{x} \overset{LH}{=} \lim\limits_{x \to 0} \frac{\left(\frac{2[e^x + 1]}{e^x + x}\right)}{1} = \frac{2(e^0 + 1)}{e^0 + 0} = \frac{2(1 + 1)}{1} = 2 \cdot 2 = 4$.

That is, $\lim\limits_{x \to 0} \ln(y) = 4$.

Now $y = y \Leftrightarrow \ln(y) = \ln(y) \Leftrightarrow e^{\ln(y)} = y$. Therefore, $\lim\limits_{x \to 0}(e^x + x)^{\frac{2}{x}} = \lim\limits_{x \to 0} y =$

$= \lim\limits_{x \to 0} e^{\ln(y)} = e^{\lim\limits_{x \to 0} \ln(y)} = e^4$.

49. Let $y = (1 + hx)^{\frac{1}{h}}$.

Hence, $\ln(y) = \ln\left[(1 + hx)^{\frac{1}{h}}\right] = \frac{1}{h}\ln(1 + hx) = \frac{\ln(1 + hx)}{h}$.

So, $\lim\limits_{h \to 0} \ln(y) = \lim\limits_{h \to 0} \frac{\ln(1 + hx)}{h} \overset{LH}{=} \lim\limits_{h \to 0} \frac{\left(\frac{x}{1 + hx}\right)}{1} = \frac{\left(\frac{x}{1 + 0}\right)}{1} = \frac{x}{1 + 0} = x$.

That is, $\lim\limits_{h \to 0} \ln(y) = x$.

Now $y = y \Leftrightarrow \ln(y) = \ln(y) \Leftrightarrow e^{\ln(y)} = y$. Therefore, $\lim\limits_{h \to 0}(1 + hx)^{\frac{1}{h}} = \lim\limits_{h \to 0} y =$

$= \lim\limits_{h \to 0} e^{\ln(y)} = e^{\lim\limits_{h \to 0} \ln(y)} = e^x$.

50. Let $y = \left(1 + \frac{x}{n}\right)^n$.

Hence, $\ln(y) = \ln\left[\left(1 + \frac{x}{n}\right)^n\right] = n\ln\left(1 + \frac{x}{n}\right) = \dfrac{\ln\left(1 + \frac{x}{n}\right)}{\frac{1}{n}}$.

So, $\displaystyle\lim_{n\to\infty}\ln(y) = \lim_{n\to\infty}\dfrac{\ln\left(1+\frac{x}{n}\right)}{\frac{1}{n}} = \lim_{n\to\infty}\dfrac{\ln\left(\frac{n+x}{n}\right)}{\frac{1}{n}} = \lim_{n\to\infty}\dfrac{\ln(n+x)-\ln(n)}{n^{-1}} \overset{LH}{=} \lim_{n\to\infty}\dfrac{\left(\frac{1}{n+x}-\frac{1}{n}\right)}{-n^{-2}} = \lim_{n\to\infty}\dfrac{\left(\frac{n-n-x}{n(n+x)}\right)}{\left(\frac{-1}{n^2}\right)} =$

$= \displaystyle\lim_{n\to\infty}\dfrac{-x}{n(n+x)}\cdot\dfrac{n^2}{-1} = \lim_{n\to\infty}\dfrac{nx}{n+x} = \lim_{n\to\infty}\dfrac{nx}{n\left(1+\frac{x}{n}\right)} = \lim_{n\to\infty}\dfrac{x}{\left(1+\frac{x}{n}\right)} = \dfrac{x}{1+0} = x.$

That is, $\displaystyle\lim_{n\to\infty}\ln(y) = x$.

Now $y = y \Leftrightarrow \ln(y) = \ln(y) \Leftrightarrow e^{\ln(y)} = y$. Therefore, $\displaystyle\lim_{n\to\infty}\left(1 + \frac{x}{n}\right)^n = \lim_{n\to\infty} y =$

$= \displaystyle\lim_{n\to\infty} e^{\ln(y)} = e^{\lim_{n\to\infty}\ln(y)} = e^x.$

51. Let $y = \left(1 + \frac{r}{n}\right)^{nt}$.

Hence, $\ln(y) = \ln\left[\left(1 + \frac{r}{n}\right)^{nt}\right] = nt\ln\left(1 + \frac{r}{n}\right) = \dfrac{t\ln\left(1 + \frac{r}{n}\right)}{\frac{1}{n}} = \dfrac{t\ln\left(\frac{n+r}{n}\right)}{\frac{1}{n}} =$

$= \dfrac{t\ln(n+r) - t\ln(n)}{\frac{1}{n}}.$

So, $\displaystyle\lim_{n\to\infty}\ln(y) = \lim_{n\to\infty}\dfrac{t\ln(n+r)-t\ln(n)}{\frac{1}{n}} = \lim_{n\to\infty}\dfrac{t\ln(n+r)-t\ln(n)}{n^{-1}} \overset{LH}{=} \lim_{n\to\infty}\dfrac{\frac{t}{n+r}-\frac{t}{n}}{-n^{-2}} = \lim_{n\to\infty}\dfrac{\frac{t}{n+r}-\frac{t}{n}}{\frac{-1}{n^2}} =$

$= \displaystyle\lim_{n\to\infty}\left(\dfrac{-tn^2}{n+r} + \dfrac{tn^2}{n}\right) = \lim_{n\to\infty}\dfrac{-tn^3 + tn^3 + rtn^2}{n(n+r)} = \lim_{n\to\infty}\dfrac{rtn}{n+r} = \lim_{n\to\infty}\dfrac{rtn}{n\left(1+\frac{r}{n}\right)} =$

$= \displaystyle\lim_{n\to\infty}\dfrac{rt}{1+\frac{r}{n}} = \dfrac{rt}{1+0} = rt.$

That is, $\displaystyle\lim_{n\to\infty}\ln(y) = rt$.

Now $y = y \Leftrightarrow \ln(y) = \ln(y) \Leftrightarrow e^{\ln(y)} = y$. Therefore, $\displaystyle\lim_{n\to\infty}\left(1 + \frac{r}{n}\right)^{nt} = \lim_{n\to\infty} y =$

$= \displaystyle\lim_{n\to\infty} e^{\ln(y)} = e^{\lim_{n\to\infty}\ln(y)} = e^{rt}.$

PROBLEMS IN RANDOM ORDER

Find the following limits.

1. $\displaystyle\lim_{x\to 0^+} \frac{\ln(x^2+2x)}{\ln(x)}$

2. $\displaystyle\lim_{x\to 0} \frac{x\cdot 2^x}{2^x-1}$

3. $\displaystyle\lim_{x\to 0} \frac{10^x-2^x}{6x}$

4. $\displaystyle\lim_{x\to\infty} \frac{e^x+x^2}{e^x-x}$

5. $\displaystyle\lim_{x\to 0} \frac{e^x+e^{-x}-2}{1-\cos(2x)}$

6. $\displaystyle\lim_{x\to 0} \frac{x^2}{\ln(\cos(x))}$

7. $\displaystyle\lim_{x\to\infty} \frac{\ln(x)}{\sqrt{x}}$

8. $\displaystyle\lim_{x\to\infty} \frac{\ln(x)}{\ln(x+1)}$

9. $\displaystyle\lim_{x\to 0^+} x\ln(x)$

10. $\displaystyle\lim_{x\to 0} \left(\frac{1}{e^x-1} - \frac{1}{x}\right)$

11. $\displaystyle\lim_{x\to 0} \left(\frac{1}{x} - \frac{1}{\sin(x)}\right)$

12. $\displaystyle\lim_{n\to\infty} \left(1+\frac{r}{n}\right)^{nt}$

13. $\displaystyle\lim_{x\to 0}(1+x)^{\frac{1}{x}}$

14. $\displaystyle\lim_{x\to\frac{\pi}{2}} \tan(x)\ln(\sin(x))$

15. $\displaystyle\lim_{x\to\infty}\left(\sqrt{x^2+3x}-x\right)$

16. $\displaystyle\lim_{x\to\frac{\pi}{2}}(2x-\pi)\sec(x)$

17. $\displaystyle\lim_{x\to 0^+}(1+\sin(4x))^{\cot(x)}$

18. $\displaystyle\lim_{x\to 0} \frac{1-\cos(x)}{\sin(x)-x+x^2}$

19. $\displaystyle\lim_{\theta\to\infty} x^2 e^{-x}$

20. $\displaystyle\lim_{x\to 1} \frac{1-x+\ln(x)}{1+\cos(\pi x)}$

21. $\displaystyle\lim_{x\to\frac{\pi}{2}^-}(\sec(x)-\tan(x))$

22. $\displaystyle\lim_{x\to\infty} \frac{e^{3x}}{x^2}$

23. $\displaystyle\lim_{x\to\infty}\left(\sqrt{x^2+x}-\sqrt{x^2-x}\right)$

24. $\displaystyle\lim_{\theta\to 0} \frac{\sin(\theta)}{\theta}$

25. $\displaystyle\lim_{x\to 0} \frac{\sin(x)-x}{x^3}$

26. $\displaystyle\lim_{x\to 0} x\cot(x)$

27. $\displaystyle\lim_{x\to\infty}(1+2x)^{\frac{1}{2\ln(x)}}$

28. $\displaystyle\lim_{x\to 0^+} x^2\ln(x)$

29. $\displaystyle\lim_{x\to 0^+} x^x$

30. $\displaystyle\lim_{\theta\to 0} \frac{1-\cos(\theta)}{\theta}$

31. $\displaystyle\lim_{x\to 0^+} x^{\frac{-1}{\ln(x)}}$

32. $\displaystyle\lim_{n\to\infty}\left(1+\frac{x}{n}\right)^n$

33. $\displaystyle\lim_{x\to\infty}\left(\frac{x^2}{x-1} - \frac{x^2}{x+1}\right)$

34. $\displaystyle\lim_{x\to 0}(e^x+x)^{\frac{2}{x}}$

35. $\displaystyle\lim_{x\to 0} \frac{2\cos(\theta)-2}{e^\theta-\theta-1}$

36. $\displaystyle\lim_{t\to 0} \frac{1-\cos(t)}{t^2}$

37. $\displaystyle\lim_{\theta\to 0} \frac{\sin(\theta)}{\theta}$

38. $\displaystyle\lim_{x\to 0} \frac{\tan(2x)}{\ln(1+x)}$

39. $\displaystyle\lim_{x\to 1^+} x^{\frac{1}{1-x}}$

40. $\displaystyle\lim_{x\to 1^+}\left(\frac{1}{x-1} - \frac{1}{\ln(x)}\right)$

41. $\displaystyle\lim_{h\to 0}(1+hx)^{\frac{1}{h}}$

42. $\displaystyle\lim_{x\to\infty}[\ln(x)]^{\frac{1}{x}}$

43. $\displaystyle\lim_{x\to\infty} \frac{\ln(x)}{x^{10}}$

44. $\displaystyle\lim_{x\to\frac{\pi}{2}} \frac{4\tan(x)}{1+\sec(x)}$

45. $\displaystyle\lim_{x\to 0} \frac{\cos(x)+2x-1}{3x}$

46. $\displaystyle\lim_{x\to 0} \frac{\sin(x)-2x}{x}$

47. $\displaystyle\lim_{t\to 0^+} \frac{\ln(t)}{\csc(t)}$

48. $\displaystyle\lim_{x\to 0} \frac{e^x-1}{\sin(2x)}$

49. $\displaystyle\lim_{x\to 0^+}\sin(x)\ln(\sin(x))$

50. $\displaystyle\lim_{x\to 1}\left(\frac{x}{x-1} - \frac{1}{\ln(x)}\right)$

51. $\displaystyle\lim_{x\to\infty}[\ln(2x)-\ln(x+1)]$

ANSWERS

1. 1 (#21)

2. $\frac{1}{\ln(2)}$ (#20)

3. $\frac{\ln(5)}{6}$ (#18)

4. 1 (#22)

5. $\frac{1}{2}$ (#11)

6. -2 (#19)

7. 0 (#13)

8. 1 (#14)

9. 0 (#25)

10. $\frac{-1}{2}$ (#34)

11. 0 (#36)

12. e^{rt} (#51)

13. e (#43)

14. 0 (#28)

15. $\frac{3}{2}$ (#37)

16. -2 (#27)

17. e^4 (#41)

18. $\frac{1}{2}$ (#15)

19. 0 (#24)

20. $\frac{-1}{\pi^2}$ (#5)

21. 0 (#33)

22. ∞ (#1)

23. 1 (#40)

24. 1 (#6)

25. $\frac{-1}{6}$ (#3)

26. 1 (#31)

27. $e^{\frac{1}{2}}$ (#47)

28. 0 (#26)

29. 1 (#42)

30. 0 (#7)

31. $\frac{1}{e}$ (#46)

32. e^x (#50)

33. 2 (#39)

34. e^4 (#48)

35. -2 (#23)

36. $\frac{1}{2}$ (#8)

37. 1 (#6)

38. 2 (#2)

39. $\frac{1}{e}$ (#44)

40. $\frac{-1}{2}$ (#38)

41. e^x (#49)

42. 1 (#45)

43. 0 (#17)

44. 4 (#12)

45. $\frac{2}{3}$ (#10)

46. -1 (#16)

47. 0 (#9)

48. $\frac{1}{2}$ (#4)

49. 0 (#30)

50. $\frac{1}{2}$ (#35)

51. $\ln(2)$ (#32)

SOME BACKGROUND MATHEMATICAL FACTS USEFUL IN CALCULATING LIMITS USING L'HÔPITAL'S RULE

$e^0 = 1$ $\ln(1) = 0$ $\ln(e) = 1$

$\sin(0) = 0$ $\cos(0) = 1$ $\tan(0) = 0$

$\sin\left(\frac{\pi}{2}\right) = 1$ $\cos\left(\frac{\pi}{2}\right) = 0$ $\cot\left(\frac{\pi}{2}\right) = 0$

$\sin(\pi) = 0$ $\cos(\pi) = -1$ $\tan(\pi) = 0$

$$\lim_{x \to \infty} \frac{1}{x} = 0 \qquad a = \frac{1}{\left(\frac{1}{a}\right)}$$

$$\ln(ab) = \ln(a) + \ln(b) \qquad \ln(a^c) = c \ln(a)$$

$$\ln\left(\frac{a}{b}\right) = \ln(a) - \ln(b) \qquad e^{\ln(x)} = x$$

$$\tan(x) = \frac{\sin(x)}{\cos(x)} \qquad \cot(x) = \frac{\cos(x)}{\sin(x)}$$

$$\sec(x) = \frac{1}{\cos(x)} \qquad \csc(x) = \frac{1}{\sin(x)} \qquad \cot(x) = \frac{1}{\tan(x)}$$

$[e^x]' = e^x$ $[\sin(x)]' = \cos(x)$ $[\sec(x)]' = \sec(x)\tan(x)$

$[\ln(x)]' = \frac{1}{x}$ $[\cos(x)]' = -\sin(x)$ $[\csc(x)]' = -\csc(x)\cot(x)$

$[a^x]' = a^x \ln(a)$ $[\tan(x)]' = \sec^2(x)$ $[\cot(x)]' = -\csc^2(x)$

TWO APPLICATIONS OF L'HÔPITAL'S RULE TO COMPARING
THE RATES OF GROWTH OF THREE FUNCTIONS

Theorem 1: $\lim\limits_{x\to\infty} \dfrac{e^x}{x^n} = \infty$ for any integer n.

Preliminary Remark: This theorem shows that the exponential function approaches infinity faster than any power of x.

Proof: $\lim\limits_{x\to\infty} \dfrac{e^x}{x^n} \overset{LH}{=} \lim\limits_{x\to\infty} \dfrac{e^x}{nx^{n-1}} \overset{LH}{=} \lim\limits_{x\to\infty} \dfrac{e^x}{n(n-1)x^{n-2}} \overset{LH}{=} \lim\limits_{x\to\infty} \dfrac{e^x}{n(n-1)(n-2)x^{n-3}} \overset{LH}{=} \cdots \overset{LH}{=} \lim\limits_{x\to\infty} \dfrac{e^x}{n!} = \infty.$

∎

Theorem 2: $\lim\limits_{x\to\infty} \dfrac{\ln(x)}{x^p} = 0$ for any number $p > 0$.

Preliminary Remark: This theorem shows that the logarithmic function approaches infinity more slowly than any power of x.

Proof: $\lim\limits_{x\to\infty} \dfrac{\ln(x)}{x^p} \overset{LH}{=} \lim\limits_{x\to\infty} \dfrac{\frac{1}{x}}{px^{p-1}} = \lim\limits_{x\to\infty} \dfrac{1}{x} \cdot \dfrac{1}{px^{p-1}} = \lim\limits_{x\to\infty} \dfrac{1}{px^p} = 0.$

∎